BACKFLOW PROTECTION FOR RESIDENTIAL SPRINKLER SYSTEMS

By:
Frederick L. Hart, P.E., Ph.D.
Robert Till
Christine Nardini
Daniel Bisson

File No. 12.054
June, 1993

Notice

This report was prepared for the U.S. Fire Administration, Federal Emergency Management Agency, Grant No. EMW-93-G-4191. The Project Officer for this study was Mr. Larry Maruskin of the U.S.F.A.. The statements and conclusions contained in this report arc those of the authors and do not necessarily reflect the views of the USFA or the Federal Emergency Management Agency.

Table of Contents

Appendices

List of Figures

List of Tables

EXECUTIVE SUMMARY

This report addresses risk vs benefit issues associated with the installation of residential fire sprinkler systems. A primary objective of this assessment is to rate the risk of potable water contamination from a residential sprinkler system, and consequently to evaluate the need for installing backflow prevention devices.

Statistics and assumptions outlined in this report show that the present risk from fire (death and/or injury) is 11.1 times higher than the present risk of waterborne disease (illness). This finding alone demonstrates a significant benefit from installing residential sprinkler systems.

The increased risk of water contamination from unprotected household sprinklers was assumed to be directly proportional to the increased volume of stagnant water added to household water lines. This calculation assumes that potable water remaining stagnant for very long periods (e.g. sprinkler lines) has the same potential for biological, chemical, and physical deterioration as water remaining stagnant for intermittent periods (e.g. typical household plumbing lines). Although equating extended stagnation to intermittent stagnation could not be supported from field or laboratory data, a number of reasons support that conclusion. In addition, because intermittent stagnation may even be more detrimental to water quality, a direct volume correlation to increased risk may actually result in a conservative risk assessment of sprinkler lines.

Reported failure rates of backflow preventers recommended for low hazard and high hazard cross connections (Double Check Valve Assemblies - DCVA and Reduced Pressure Principle Backflow Devices - RPBD) were used to estimate the reduction of water contamination risk when using these devices. An estimated failure rate of a simple and low cost alternative device (Single Check Valve - SCV) was used to predict the reduction of water contamination risk when using that device.

Four installation options were selected:

Scenario A: No protection device (SCV, DCVA, RPBD) is provided

Scenario B: Either of the above three devices (SCV, DCVA, RBBD) are placed at the sprinkler line.

Scenario C: Either of the above three devices (SCV, DCVA, RPBD) are placed at the sprinkler feed line and also at the potable water feed line.

Scenario D: Same as Scenario C, but the device is placed at the feed line before branching to the potable feed line and the sprinkler feed line.

A normalized risk (death & injury from fire per population base, and waterborne illness per population base) was calculated and presented as follows:

Condition	Fire Event (Injury & Death)	Waterborne Event (Illness)
Household Water Only (No Sprinkler)	11.10	1.00
Scenario A	5.84	2.50
Scenario B	5.84	
SCV		1.06
DCVA		1.02
RPBD		1.03
Scenarios C & D	5.84	
SCV		0.10
DCVA		0.04
RPBD		0.05

Normalized Risk of Fire (Injury & Death) and Waterborne Illness Events

Based on this relative comparison of risk, and also noting that at no time did the risk of waterborne illness reach the USEPA acceptable level of risk for gastrointerstinal illness (1×10^{-4} or a relative risk of 12.2), the following recommendations are made:

1) The SCV device should be considered in favor of the DCVA and the RPBD for individual household sprinkler systems.

2) If a protection device is to be used for a sprinkler line, the same degree of protection should be considered at the potable water line.

INTRODUCTION

An estimated 478,000 residential fires occurred in the United States in 1991 accounting for 74.6% of all the structure fires in that year. Of these residential fires, one- and two-family dwellings accounted for 363,000 which represents 56.7% of all structure fires (1). In this one year, fires that occurred in one- and two-family dwellings are estimated for causing 2,905 deaths (65% of total deaths), 15,600 serious injuries (53.1% of total injuries) and $3,354,000,000 in property loss (40.1% of total property loss).

Much of this suffering and financial loss could have been substantially reduced if residences **were** equipped with sprinkler systems. In a study done by Ruegg & Fuller, it was estimated from simulation that sprinklers in dwellings can reduce the number of fire deaths by 63% and the number of non-fatal casualties by 44% (2). These simulations involve scenario techniques which use a benefit-cost model to predict the effectiveness of sprinkler systems in residential homes.

Equipping houses with sprinkler systems without a check valve, however, may lead to another problem; public water supply contamination. This potential exists because sprinkler pipe lines will contain stagnant water which, under certain conditions, could be reintroduced back into the public water network. In 1991, Regli, et. al. (3) reported that over 500 outbreaks (multiple illnesses in the same distribution system or area) of waterborne disease occurred since 1971. In a 1982 report (4). EPA reported 27,000 cases of illness from the consumption of contaminated water during the period 1972 through 1976 and reported an additional 11,435 cases of waterborne illness in 1987. In a more recent study, 26 outbreaks of waterborne disease were reported for the years 1989 and 1990. Of the reported outbreaks for this two year time frame, 11 of the outbreaks were attributable to community water systems which resulted in 1,660 illnesses and four deaths (5). It should be noted, that none of the above mentioned cases were attributed to backflow failure from sprinkler systems.

The potential for water contamination can be reduced by isolating sprinkler lines from the public water supply with backflow preventers. A simple and low cost version of a backflow preventer is the single check valve (SCV), which has been used on sprinkler systems since the early 1800's. Many agencies concerned with water quality issues, however favor a double check valve assembly or a reduced principle backflow preventer device (RPBD) because they feel that this added degree of protection is warranted. These more sophisticated devices, however, are expensive and may deter people from installing household sprinkler systems. The RPBD's may also negatively impact the usefulness of a sprinkler system by reducing the available pressure because of friction losses, and therefore lowering the deliverable flow. Finally, regulatory requirements often associated with sophisticated backflow preventers might also detract home owners from considering a sprinkler system.

Consequently, two equally significant public health and safety issues (public safety through fire protection and public safety through water supply protection) are at direct odds with each other. This situation is unavoidable for most household dwellings because homes typically rely on the same water source to satisfy both objectives. These opposing viewpoints cause advocates on each side to often consider the other safety issue as a direct threat to their own concerns. For example, a requirement for sprinkler systems in individual households would satisfy fin safety concerns but, without the presence of backflow preventers, potable water supplies may be jeopardized due to the introduction of stagnant water into potable water lines. Conversely, the requirement for certain types of backflow preventers on sprinkler lines may make the entire household sprinkler system prohibitively expensive, and for retrofit systems - possibly incapable of functioning properly.

These opposing views have existed for a number of years and have been debated extensively, often under highly emotional circumstances. Now that the need for residential sprinkler systems are being seriously recognized (especially for new home construction), the debate has intensified. Typically, those concerned with water quality issues favor a maximum degree of protection provided by a double check valve assembly or a reduced principle backflow device, while those concerned with fire protection issues favor a less expensive, low head loss single check valve device. Others may even favor no backflow preventers for individual households because of the regulations that are often associated with the installation and inspection of backflow devices.

This paper addresses this debate by focusing on the following question:

What is the difference in risk associated with a single check valve &vice compared to a double check valve assembly and a reduced principle backflow &vice for residential sprinkler system?

To answer this question, an evaluation of risks and benefits associated with either decision is presented. For both options. the following assumptions are made:

- All water in a household plumbing system (including sprinkler system) is from a community water system.

- The household is supplied with high quality municipal water that meets drinking water requirements at the point of delivery.

- The sprinkler pipeline is constructed of either copper (no lead solder) or PVC pipe suitable for potable water supplies.

- The probability of fire, backflow device failure and waterborne disease outbreak are typical and arc based on data reported in the literature.

This paper is subdivided into the following sections:

Section 1.0: Definitions
A list of terms used in this report along with their definations are presented.

Section 2.0: Background on Current Standards and Regulations
This section identifies agencies and authorities that are associated with fire protection and public water supply and identifies their concerns relative to household sprinkler systems and water quality issues.

Section 3.0: Typical Household Piping Systems
This section presents typical components of a household water supply pipe system and a typical sprinkler pipe system.

Section 4.0: Possible Water Quality Changes
Water quality changes (biological. chemical and physical) reported to occur during normal household water flow conditions and stagnant conditions are presented.

Section 5.0:   Available Data

Background data on fire incidents, disease outbreaks and backflow preventer failures are presented.

Section 6.0:   Risks versus Benefits

A risk versus benefit evaluation is presented for cases where various backflow preventers are present and conditions of fire or water contamination occur.

Section 7.0: Areas Needing Additional Information

During this literature survey, some information was not available. This information, which either does not exist or was not found by the authors, is identified here.

Section 8.0: Summary and Conclusions

Based on information gathered in this study, a summary of risks and benefits associated with the use of single check valve, a double check valve assembly, and a reduced principle backflow preventer is presented. Finally a list of recommendations are given.

SECTION 1.0: DEFINITIONS

Backflow Prevenier - A mechanical device which prevents a reverse flow of water (SCV, DCVA, RPBD).

Single Check Value (SCV) - A valve similar in reliability to a check valve used for a DCVA, which allows flow in only one direction, from the supply to the distribution system. This type of device is not normally referred to as a backflow preventer. However, because this report compares the SCV with more expensive backflow preventers, a SCV is included in the backflow preventer definition.

Double Check Valve Assembly (DCVA) - A mechanical device which consists of two independently acting check valves with test cocks and shut-offs at both ends. This assembly is typically specified as a backflow device at low hazard cross connections.

Reduced-Pressure Principle Backflow-Prevention Assembly (RPBD) - **A** backflow prevention device that consists of two independently acting, check valves together with a mechanically independent pressure differential relief valve located between the check valves and below the first check valve. These units are located between two tightly closing resilient-seated shutoff valves as an assembly, and are quipped with properly located resilient-seated test cocks. This assembly is typically specified as a backflow device at high hazard cross connections.

Cross-Connection - Any unprotected actual or potential connection or structural arrangement between a public or a consumer's potable water system and any other source or system through which it is possible to introduce into any part of the potable system any used water, industrial fluid, gas or substance other than the intended potable water with which the system is supplied.

Residential Distribution System - A piping system which begins at the water service line connection and distributes water throughout the house for domestic use.

Residential Sprinkler System - A piping system within the household that begins at the water service line connection and distributes water to the fire sprinklers.

Potable Water - Water from any source which has been investigated by the health agency having jurisdiction, and which has been approved for human consumption.

Community Water System - Publicly owned or investor-owned systems that serve large or small communities, subdivisions or trailer parks with at least 15 service connections or 25 year-round residents.

Biological Contamination - The introduction of biological organisms into the potable water supply which is dangerous to the public's health because it may cause sickness, disease and possibly death.

Chemical Contamination - The introduction of toxic substances into the potable water supply which is dangerous to the public's health because of poisoning which may cause sickness or death.

Physical Contamination - The introduction of turbidity or color causing materials into the potable water supply which may contribute to biological and/or chemical contamination.

Biofilms - Organic or inorganic deposits consisting of microorganisms, microbial products and detritus at a surface (6).

Sloughing - The detachment of macroscopic patches of biofilm from a surface (pipe) (7).

Assimilable Organic Carbon (AOC) - The portion of total organic carbon (TOC) that can be readily used by aquatic organisms for growth (8).

Risk Level - The U.S. EPA adopted acceptable level of risk for gastrointestinal illness is one illness Per 10,000 People per year (10-4 annual risk) (9).

Outbreak of Waterborne Disease - An incidence when at least two people experience a similar illness after ingesting or using water intended for drinking. Epidemiologic evidence must implicate the water as the source of illness (5).

Stagnation - Zero to very low water velocity conditions. Intermittent stagnation is assumed to occur in normal household plumbing while extended stagnation is assumed to occur in a sprinkler system.

SECTION 2.0: REGULATIONS AND AUTHORITIES

There are a number of regulatory agencies and authorities associated with drinking water standards, cross connections to potable water supplies, backflow preventers and sprinkler systems. This section identifies and briefly outlines their relevance to the application of household sprinkler systems.

Water Quality

The United States Environmental Protection Agency (USEPA) was mandated by the Safe Water Drinking Act (SWDA), and later by its amendments, to establish potable drinking water standards. These standards are now adopted, as minimum requirements, by local authorities which have jurisdiction **over** public water suppliers. The 1986 SWDA Amendments have a direct relevance to sprinkler systems as they refer to water quality at the point of use in addition to the point of delivery from the water treatment plant. Therefore, the potential for contamination within the distribution systems has become a direct concern to water quality standards. These standards include new coliform regulations, turbidity regulations and copper regulations.

The new coliform rule, which became effective on June 29, 1989, established a maximum contaminant level (MCL) goal based on the presence or absence of total coliforms, whereas, the original standard was based on the bacterial density. Coliform occurrences caused by bacterial growth in the distribution systems may cause violation of the new coliform rule even though a demonstrable public health risk may not exist The potential for colifotrn occurrences from the re-introduction of stagnant water within the distribution system, therefore, may be of significant concern to water quality standards. The USEPA does, however. allow variances for systems that are not at risk for fecal or pathogenic contamination (9).

The lead and copper rule became effective in June 1991. This new rule applies to all water systems. Instead of setting MCL's for lead and copper, this new regulation specifies action requirements consisting of corrosion control and public education. The action level for copper is 1.3 mg/L. Because water stagnating in copper pipelines may contribute to elevated levels of copper, sprinkler pipelines may pose a concern with regard to this new water quality standard (10).

Turbidity particles have been classified for years as a secondary standard. A secondary standard pertains to contaminants that adversely affect the aesthetic quality of water. With the advent of the Surface Water Treatment Rule (SWTR), which became effective on December 30, 1990, turbidity was classified as a primary standard. A primary standard relates to public health protection. This new rule sets a turbidity limit of 5 NTU, and typically requires 95% of all samples to have a turbidity of 1 NTU or lower. Turbidity particles, although not considered harmful in themselves, may harbor microorganism and chemicals which may be harmful. In addition, turbidity could contribute to corrosion reactions. Unlike the copper and coliform rules, turbidity standards are concerned with water leaving the water treatment plant rather than within the distribution system. Nevertheless, increases of turbidity within the water supply from stagnant water lines should be viewed as a problem from the public health perspective.

Cross Connection Authorities and Regulations

Individual state environmental agencies have :he highest form of authority in each state for regulating cross connections policy and the need for backflow preventers. The agencies mandate that water suppliers submit a cross connection program plan. In Massachusetts, for example, the plan is defined by the Drinking Water Regulations of Massachusetts (310 CMR 22.22) and enforced by the Department of Environmental Protection (DEP). It states that a plan shall be submitted by each water supplier describing the current and proposed cross connection programs and include information on staff training, testing, surveying, fee structure, and other areas.

Every supplier of water is responsible for the safety of the public water system under its jurisdiction. They must inspect and survey all areas of a public water system to find if cross connections exist, record all findings, take appropriate actions to alleviate possible hazardous conditions and report yearly to the DEP. Some suppliers of water for the state of Massachusetts are delegated authority to act under code 310 CMR 22.22 and thus are a Department Designee. They have the responsibility to approve plans for new installations of back flow preventer devices and to inspect the new installations.

The DEP requires any new or existing fire system that is connected to street-water mains to be evaluated by the Department or its Designee for the determination of what type of backflow preventer is required. This may include residential sprinkler systems. According to 310 CMR, however, residential sprinkler systems that do not have any fire pumper connections, nor any chemical additives and are made with potable piping material would not require a backflow preventer. Finally, sprinkler system connections that originate within a house are not regulated by DEP and subsequently are under the jurisdiction of local plumbing codes.

Codes Relevant to Backflow Preventers

Regulatory agencies will refer to plumbing codes and manuals published by other agencies when specifying rules and regulations concerning backflow preventers. The following plumbing codes are commonly used as references:

1. *The National Plumbing Code* (NPC)
   - Building Officials and Code Administrators International, Inc.

2. *The Uniform Plumbing Code* (UPC)
   - International Association of Plumbing and Mechanical officials

3. *The Standard Plumbing Code* (SPC)
   - Southern Building Code Congress International

The NPC is primarily based on the AWA M-14 manual (39). The new 1993 NPC contains rquirements that all automatic fire sprinkler systems be protected against backflow. The code requires a double check valve assembly on all sprinkler systems. However, systems that use approved water piping materials and systems containing no fire department connections arc exceptions. Therefore, residential sprinkler systems that use approved potable piping and do not have fire department connections would not require backflow according to the NPC.

The 1991 UPC contains requirements that piping not made of approved potable piping materials be separated by an approved backflow prevention device.

In the area of fire sprinklers, the 1991 SPC specifically references the AWWA M-14 manual. The SPC will be updated to be in accordance with the revised 1990 edition of the AWWA M-14 manual

The AWWA M-14 Manual, *The Recommended Practice for Backflow Prevention and Cross Connection Control,* was revised in 1990. This manual provides classifications and guidelines for the recommended use of backflow preventers. In Chapter 3, a cross connection is defined to mean "any unprotected actual or potential connection or structural arrangement between a public or a consumer's potable water system and any other source or system through which it is possible to introduce into any part of the potable system any used water, industrial fluid, gas, or substance other than the intended potable water with which the system is supplied." From this definition, the manual classifies each case according to its degree of risk to the potable water system In Chapter 6 of the manual; it is stated that for cross connection control, fire-protection systems may be classified on the basis of water source and the arrangement of supplies. A residential sprinkler system would fall into the category of Class 1 or Class 2.

A Class 1 consists of direct connections from public water mains only; no pumps, tanks, or reservoirs; no physical connection from other water supplies; no antifreeze or other additives of any kind; all sprinkler drams discharging to atmosphere, dry wells, or other safe outlets. A Class 2 is the same as a Class 1 except that a booster pump may be installed in the connections from the street mains. For Class 1 and Class 2 fire protection systems, an approved backflow preventer is not required.

SECTION 3.0: TYPICAL HOUSEHOLD WATER PIPING SYSTEMS

Introduction

Descriptions of typical household plumbing and sprinkler systems arc presented in this section. With these piping systems clearly defined, a quantification of stagnant water associated with a sprinkler pipeline system and a typical household plumbing system is possible. The household system presented here consists of a 3/4 inch supply line that is branched at the house to the residential distribution system and the residential sprinkler systems. Both systems (potable water supply lines and sprinkler supply lines) are assumed to be constructed of copper or PVC approved for potable water use.

Potable Water Pipe Design

Design of the residential piping system was based on a typical 20 feet by 50 feet ranch style single family home. The hot and cold water piping system was designed according to plumbing codes and a plumbing design manual (11). A list of specifications used in the design can be found in Appendix 1. A schematic of the design is shown in Figure 1.

Potable Water piping for the house consisted of two different diameters: 1/2 inch and 3/4 inch. The total length of pipe contained in the system was determined to be 45 feet in length for the 1/2 inch pipe and 90 feet for the 3/4 inch pipe. Pipe lengths were estimated from the service line to the water use locations inside the house. The calculated volumes of the 1/2 inch and 3/4 inch pipes was 106 cubic inches and 477 cubic inches respectively, totalling 583 cubic inches.

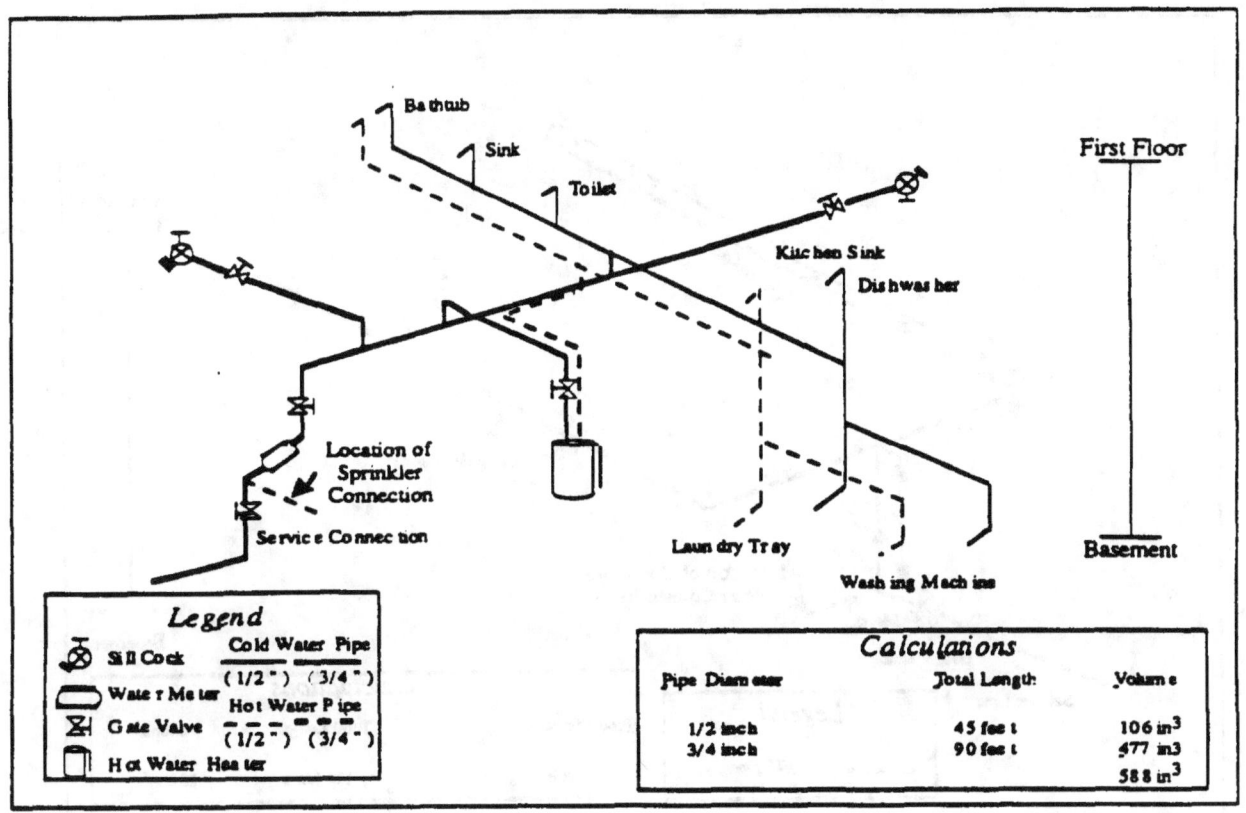

Figure 1 - Typical Hot and Cold Piping System for a Single Family Dwelling

Sprinkler System Design

The sprinkler system was designed according to the standards found in the National Fire Protection Association (NFPA) 13D and can be found in Appendix 2. A schematic of this system is shown in Figure 2.

The diameter of the service line for the sprinkler was also considered to be the standard 3/4 inch in diameter. The total length of the sprinkler piping was estimated from the service line connection to the system connection inside the house. The total length of the sprinkler pipe line was estimated from the service line connection to the system inside the house and was determined to be 113 feet. The total volume of water contained in the sprinkler line was calculated to be 896 cubic inches.

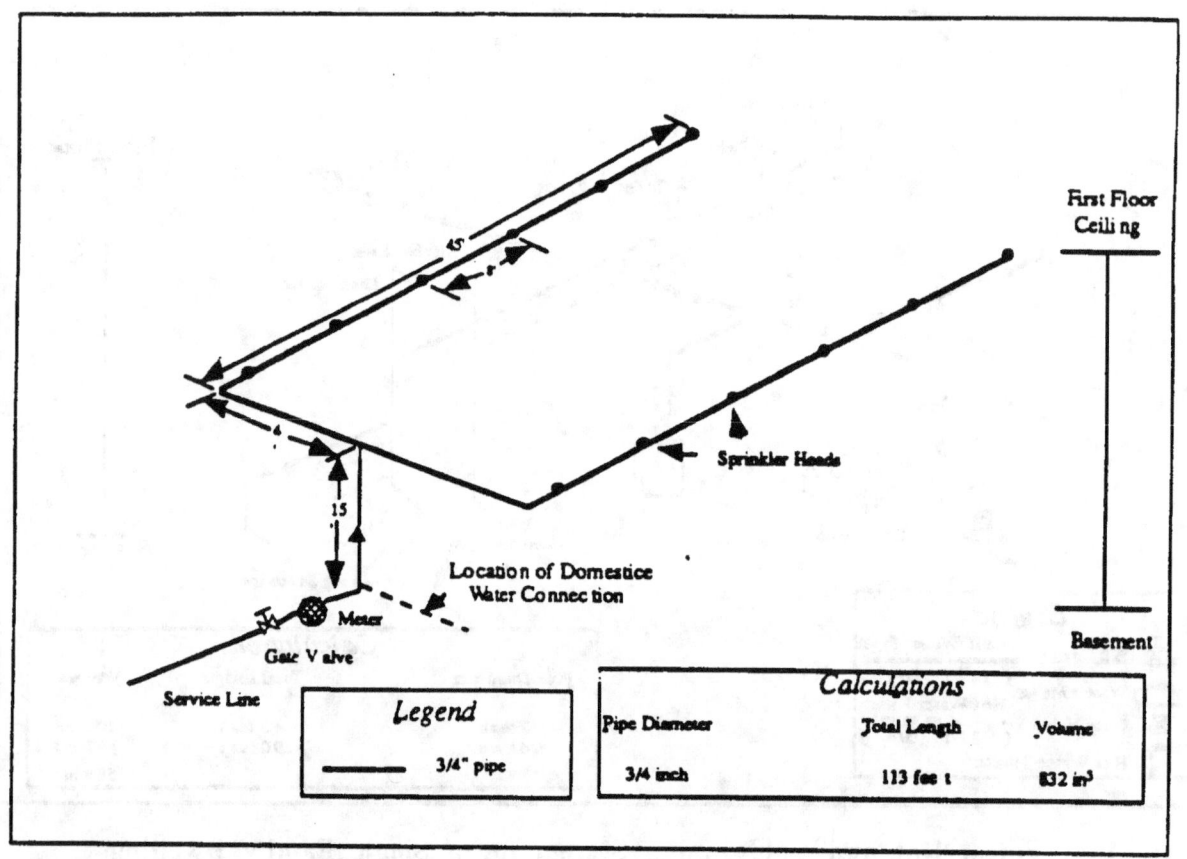

Figure 2 - Typical Sprinkler System For a Single Family Dwelling

Household Plumbing Network vs the Sprinkler Network

Based on the above design calculations, the total volume of water in a household plumbing system should be increased by 2.5 if a sprinkler system is installed. As will be pointed out later, a parallel projection of increased risk from this added volume may not be totally rationaL However, because there are a number of arguments that may be used to support suspicions that **a** relatively short term stagnant pipe system (household plumbing) represents equal or increased dangers compared to extended stagnant pipe systems (sprinkler systems), an assignment of equal risk is conservative.

16

SECTION 4.0: POSSIBLE WATER QUALITY CHANGES

Water left stagnant in a water line used for potable water delivery or for fire sprinklers will deteriorate in terms of biological, physical, and chemical parameters. This chapter discusses these changes.

Biological Quality Changes

Major factors that determine microbial water quality include break through, biofilm formation, and growth rates (6). Given the correct set of circumstances, even the most sophisticated water treatment systems can experience negative impacts on the quality of water in their distribution systems.

Breakthrough

Breakthrough is the increase in bacteria in the distribution system due to bacteria passing through the treatment system It is suspected that a significant portion of the breakthrough bacteria may be injured bacteria that is not detected using standard tests for treatment plant effluent. Furthermore, bacteria may survive disinfection and pass through the treatment system by **adhering** to particles, including macroinvertebrates, turbidity particles, algae, and carbon fines. As a rule, turbidity interferes with the detection and accurate accounting of bacteria allowing turbidity associated organisms to go undetected or under counted (6).

Biofilms

Microbial growth in distribution systems is largely attributed to the occurrence of biofilms on pipe surfaces (7). A biofilm is defined as an organic or inorganic deposit consisting of microorganisms, microbial products and detritus at a surface. Biofilm formation is initiated by the transport and accumulation of microorganisms and nutrients at an interface. followed by metabolism. growth, and product formation, and finally, detachment, erosion or sloughing of the biofilm from the surface (6). The rate of biofilm formation is dependent on the physicochemical properties of the interface, the physical roughness of the surface, and the physiological characteristics of the microorganisms (12). During a study of a New Jersey water utility, LeChavallier concluded that coliform bacteria found in water columns originated from biofilms in the distribution system and that the coliform levels increased 20-fold as the water moved through the distribution system, away from the treatment plant. Coliform species diversity also increased as the water flowed away from the treatment plant (9).

Studies have also shown that stagnation in distribution systems can lead to loss of the disinfectant residual, thus promoting microbial growth. Water in dead end lines is consistently found to be of poorer quality than water found elsewhere in the same distribution system (6). An inverse relationship has been demonstrated between water velocity and biofilm counts (6,14). Furthermore, the starting and stopping of water flow in pipe systems has been shown to increase bacterial levels up to ten-fold (6). Emde et al. reported that microbial regrowth in distribution systems is linked to the recurrent sloughing of pipe biofilms into passing water (15). LeChavallier et al. concluded from a study of a drinking water distribution system experiencing recurrent coliform bacteria growth, that the majority of water quality deterioration could be attributed to the dead-end sampling locations. This study emphasized that short periods of water stagnation can have profound and adverse effects on the quality of potable water (16).

Bacterial Growth

LeChavallier indicated that numerous factors contribute to bacterial survival in chlorinated water supply systems, including: attachment to surfaces, bacterial aggregation, age of the biofilm, encapsulation, previous growth conditions, alterations in the bacterial cell wall, and choice of disinfectant (6). In drinking water distribution systems, several conditions can promote the growth of biofilms. Macromolecules tend to accumulate at the solid- liquid interface, creating surface area and protection for the attachment of microorganisms. High flow rates can transport nutrients to fixed microorganisms, providing the necessary nutrients for growth in otherwise low nutrient concentration water. Furthermore, biofilms can provide shielding of microorganisms from disinfectant residuals. Research has shown that low-level breakthrough of injured bacteria from the treatment plant combined with increases of bacterial densities by subsequent growth in the distribution system has caused coliform occurrences. Such breakthrough and regrowths are characterized by large initial occurrences of coliform organisms followed by a gradual decline in bacterial levels over time. The period of time between the large initial occurences and the gradual decline has been observed to last several months (12).

Bacterial growth has been attributed to several conditions affecting a distribution system including: environmental factors (e.g. temperature and rainfall), nutrient concentrations, ineffectiveness of disinfectant residuals, corrosion, and hydraulic effects (12).

Microbial activity is reported to increase with temperature, with significant activity commencing at temperature of 15°C and higher. Temperature has been found to influence both the growth rate, the lag phase, and cell yield (12). In a year long study of bacterial growth in a distribution system, LeChavallier et al. found that coliform organisms occurred most frequently during warm weather periods (8). Donlan and Pipes concluded a strong relationship between water temperature and attached microbiological population density (17).

Heterotrophic bacteria require carbon, nitrogen, and phosphorous in a ratio of 100:10:1, respectively; therefore, organic carbon may often be the growth-limiting nutrient in drinking water (8). Assimilable organic carbon (AOC) is the portion of total organic carbon that is usable by microorganisms as nutrient for growth. AOC levels have been found to decline in drinking water as it flowed through the distribution system (6). LeChavallier reported the results of AOC on bacterial growth as waters containing greater than 50 pg/L showed bacterial growth. Specifically, E.coli was demonstrated to be inhibited at AOC levels of less than 54 pg/L (12).

Physical Quality Changes

Water pipes experiencing stagnation are prone to tuberculation. This phenomenon, which is a result of chemical and biological reactions, is characterized by deposits on the interior pipe wall (7). Tuberculation can severely restrict waterflow and contribute to water quality deteriorations if settled deposits are re-suspended from flow surges. Tubercule growth initiates from iron oxide blisters on ferrous pipe walls. This phenomenon, therefore, is not expected to be prevalent in copper or PVC lines.

The primary form of physical water quality deterioration is turbidity increases due to a re-suspension of deposited materials. Turbidity particles, although not necessarily harmful themselves, can contribute to bacterial growth and corrosion reactions. For potable water lines made of copper or PVC, the accumulations of turbidity causing scales or deposits should be a function of the raw water quality rather than potential tuberculation reactions. Water lines that experience intermittent stagnation (such as dead ends in potable supplies) may be more prone to resuspension occurrences than water lines experiencing extended stagnation. The degree of intermixing at the extended stagnant sprinkler line to the intermittent stagnant potable water line is not clearly understood at this time.

Chemical Water Quality Changes

Chemical water quality changes resulting from stagnation will occur from corrosion reactions. Conditions that influence these reactions include: water quality, pipe material, and flow conditions (18). Of the two materials considered in this paper (copper and PVC), only copper corrosion is feasible, as the PVC will not support the necessary electrical reactions.

Water Quality

The primary parameters that will influence corrosion are pH, oxygen levels, carbon dioxide levels, water conductivity, and temperature. Low pH conditions will prevent protective scale build-up and consequently result in uniform copper corrosion (19). This type of reaction may cause copper leaching to the water which results in a quality deterioration. The presence of oxygen and high carbon dioxide levels will promote pitting, which results in **a** corrosion reaction that causes pipe failure. Elevated conductivity and temperature conditions will increase corrosion reactions (20).

Pipe Material

Copper corrosion is low compared to other pipe materials because a protective CuO_2 film will form on the pipe wall. In the absence of oxygen, this protective film may break down, causing pitting (20). Copper pipe is generally considered to be corrosion resistant (21).

Flow Conditions

During intermittent stagnation conditions, products of corrosion may be flushed, exposing new pipe material. In addition, the extent of corrosion may be more uniform throughout the pipe line with flowing water conditions. Extended stagnation, however, will cause corrosion to be more localized (21).

Summary - Water Quality Changes

The literature does indicate that biological, physical, and chemical water quality changes will occur in water lines if stagnation or low velocity conditions exist. An examination of these phenomena, however, supports a suspicion that extended stagnation conditions may not be as detrimental to water quality as intermittent stagnation conditions. For example, a large volume of information is available that clearly shows the probability of bacterial growth in stagnant water lines. But data supporting this conclusion is based on potable water lines that are subjected to intermittent stagnation. Resupplies of needed nutrients and oxygen will be provided in that type of environment, but may not be available under extended stagnation conditions. Biofilm sloughing and nutrient replenishment should occur at local zones where sprinkler lines are attached in the absence of any flow reversal protection, but the degree of intermixing at these connections is unknown. Similarly, corrosion reactions that can occur in water lines supplied with oxygen and particles may be more prone to chemical and physical water quality deterioration conditions than water lines that very rarely (or never) experience water flow.

One could take this argument to an extreme and conclude that extended stagnant water is actually safer than intermittent water. A more conservative approach, however, would be to conclude that they have equal degrees of risk associated with them. Potable water lines within a household (intermittent stagnation), therefore, is assumed to be equal in risk as sprinkler lines within a household (extended stagnation) if both lines are fed with the same initial water and are made of PVC or copper approved for potable water delivery.

SECTION 5.0: AVAILABLE DATA

Introduction

This section presents data pertaining to fire incidents, disease outbreaks and backflow preventer failures currently available in the literature. Deaths and injuries related to fires was readily available from existing databases and is presented as historical information on occurrences in one and two-family dwellings. Information pertaining to fire-related deaths and injuries occurring in one and two-family dwellings quipped with sprinkler systems was statistically insignificant; therefore, simulated data is presented. Data pertaining to the historical occurrences of illnesses and deaths associated with waterborne disease is available, however, much of this data does not identify the source of the contamination (treatment plant failure, cross-connection, etc.). One recent study that provided waterborne disease outbreaks identified by source of contamination is presented. Data pertaining to the performance reliability of backflow preventers was based on actual field studies conducted by a large community water authority over a period of three years, representing approximately 6,000 tests.

Fire-related Deaths

The total number of civilian deaths in one and two family dwellings by year is shown below in Figure 3 (22). Since 1977-78. some trends arc worth noting. Deaths in one and two family dwellings peaked in 1978, where 5,213 occurred. Deaths then decreased steadily during the 1979-82 period to 3,969 by the end of 1982. the only exception being the year 1981.

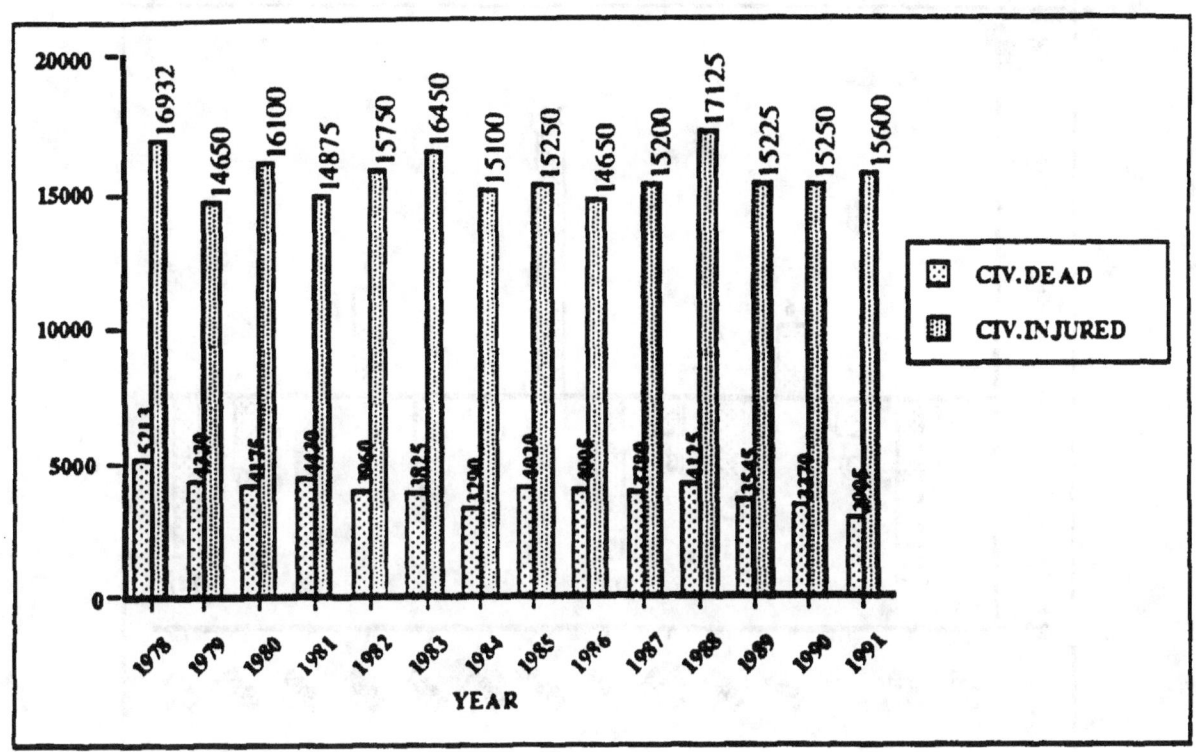

Figure 3 - Number of Civilian Deaths and Injuries for One and Two Family Dwellings by Year

From 1982 to 1988, the number of home fire deaths remained relatively constant in the 3,780 to 4,125 range. The only exception was 1984, when 3290 deaths occurred. In the past three years, home fire deaths moved well below the 1982-1988 plateau. There were 3,545 in 1989, 3,370 in 1990, and 2,905 in 1991, the last year for which information is currently available. The percent changes from the previous year are shown below in Figure 4 (22). It should be noted that 1978 data is not included, due to a change in NFPA survey methods. The year 1991 was the third year of a home fire death total that was significantly lower than the 82-88 plateau. This shows that progress in limiting the number of fire deaths in the home is currently being made, possibly due to increased use of home smoke detectors, increased public fire safety education, and increased use of more fire safe products (23).

23

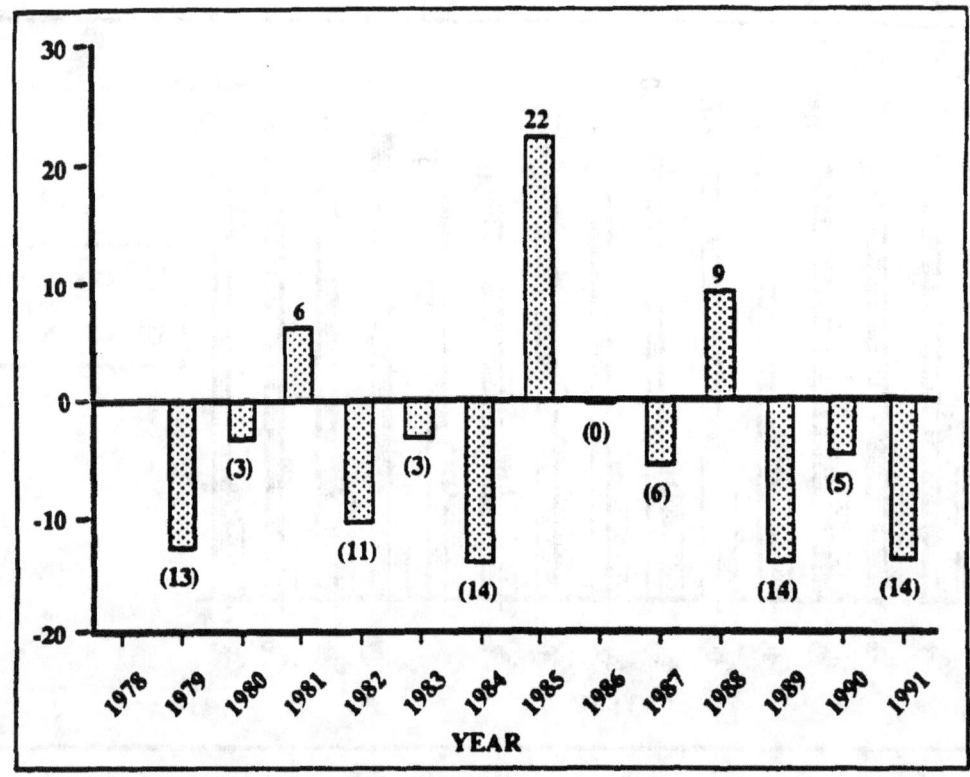

Figure 4 - Percent Change in Civilian Deaths From Previous Year

Reported Fire-related Injuries

Estimates of civilian injuries in one and two family dwellings tend to be low. because many civilian injuries are not reported to the fire service. For example, many injuries occur at small iires that fire departments do not respond to. In addition, sometimes when departments do respond they may be unaware of injured persons that were not transport to medical facilities (24).

It can be seen from the data in Figure 3 that the total number of injuries per year is in the 15,000 range, with slightly higher numbers in some years, particularly in 1978 and 1988. The per year percent changes in these numbers is shown below in Figure 5 (22). Data for 1978 is missing, due to changes in NFPA survey methods, as noted above.

24

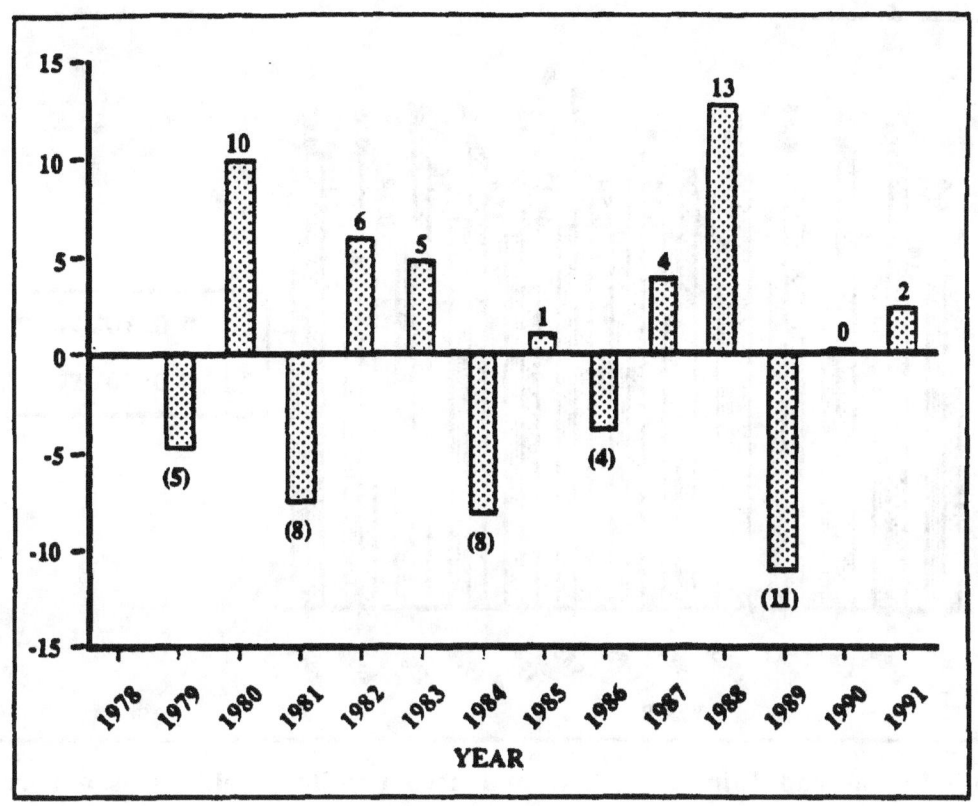

Figure 5 - Percent Change in Civilian Injuries From Previous Year

In 1991, 65 percent of all reported fin deaths occurred in the one and two family dwellings. As shown in Figure 6, consistently about two-thirds of all fire deaths occurred in one and two family dwellings. Also shown in Figure 6, consistently roughly 50 percent of all reported fin related injuries occur in one and two family dwellings.

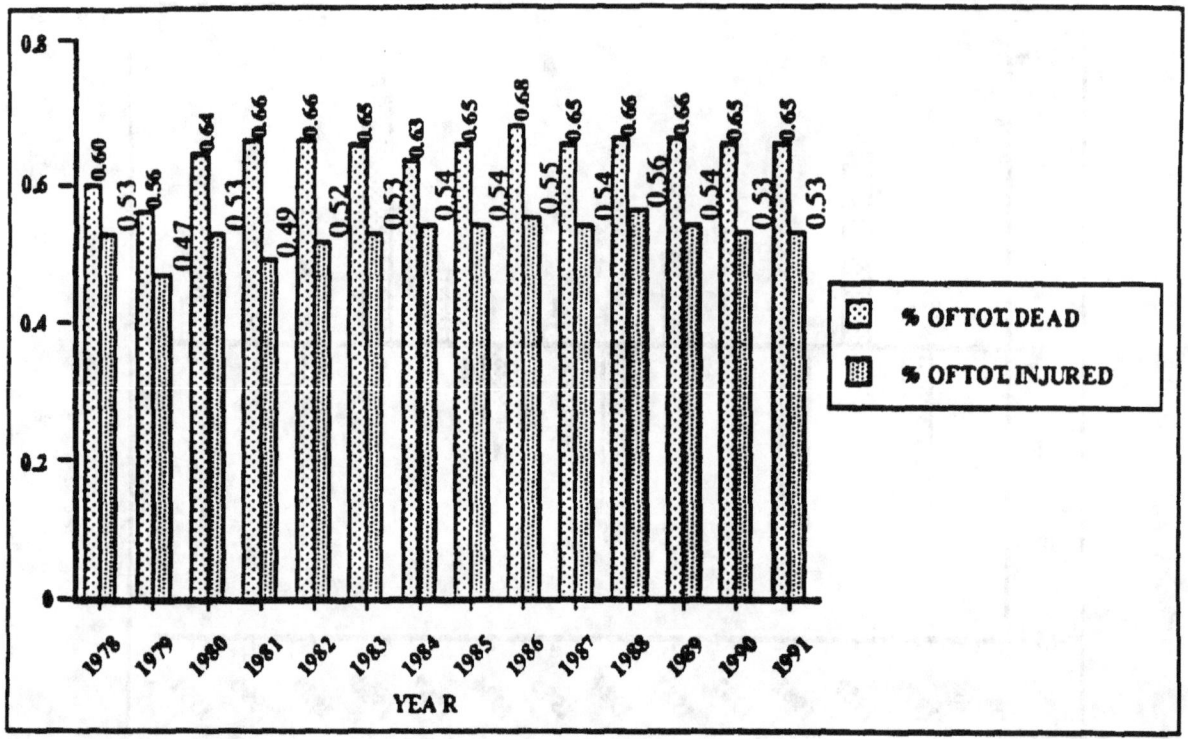

Figure 6 - Deaths and Injuries in One and Two Family Dwellings as a Percentage of Total Fire Related Deaths and Injuries

This information suggests that one and two family dwellings account for the majority of deaths and injuries attributed to all fires in the United States. The wider use of residential sprinklers have been suggested to reduce the number of fire deaths and injuries occurring in the home (24).

Data on Effects of Residential Sprinklers

A number of studies have been performed attempting to determine the effectiveness of residential sprinkler systems. These include studies that apply information concerning standard commercial sprinkler systems to residential occupancies, such as the study performed by Melinek (25). They also include studies using information that strictly concerns residential systems, such as those performed by Ruegg and Fuller (2.27) and Harmathy. Automatic sprinklers are thermosensitive devices designed to react at predetermined temperatures by automatically releasing a stream of water and distributing it in specified patterns and quantities over designated areas (28). Residential fast response sprinklers differ from standard response sprinklers for a number of reasons. These sprinklers have a higher thermal sensitivity than standard response sprinklers, which allows for quicker activation than standard response sprinklers. In addition to their fast response characteristics, residential sprinklers have a special water distribution pattern. This pattern must be more uniform than the pattern of standard spray sprinklers, because in large areas standard spray sprinklers can rely on the overlapping patterns of several sprinkler heads to make up for voids (28). For these reasons, and others, information concerning standard sprinkler systems as applied in residential environments, such as that produced by Mclinek (25) is not readily applicable to residential fast response sprinkler systems.

Information about 201,417 fins that occurred in 1990 in one and **two** family dwellings is available from the National Fire Incident Reporting System (NFIRS) database. Data is collected by each of the participating states and gathered together by the Federal Emergency Management Agency (FEMA). This data is coded in accordance with the appropriate NFPA standards. The NFIRS database showed that less than 0.5 percent of one and two-family dwellings arc equipped with sprinkler systems. Furthermore, it is unknown if these systems complied with NFPA 13D. Since the number of fires in sprinklered homes is not statistically significant, we must depend on the results of research studies consisting of field tests, experiments. and expert judgement, to determine the reduction of deaths and injuries that will occur due to the installation of residential sprinkler systems.

Ruegg and Fuller prepared a cost benefit analysis of residential sprinkler systems in which they adopted death and injury information from an unpublished report by Gomberg, Hall, Stiefal, Offensend, and Pacey. Gombcrg et al. conducted a study of the loss reductions offered by fast response sprinklers. Their work was planned to be published by the National Bureau of Standards (29). According to the study, sprinkler effectiveness was based on expert judgement and "extrapolations applied to the results of laboratory and field tests". The percent reduction in human losses due to sprinklering was determined in this report to be an 80.4 percent reduction in fire deaths, and a 45.9 percent reduction in fire injuries.

Ruegg/Fuller obtained results by subtracting the predicted deaths and injuries per fire assuming smoke detectors and sprinklers, from the average deaths and-injuries assuming the use of smoke detectors alone. Reliabilities of 84 percent and 92 percent for smoke detectors and sprinkler systems, respectively, were assumed (2). This information, combined with the effectiveness measures cited in the unpublished report, yielded a 63 percent decrease in life loss due to sprinkler presence, and a 44 percent decrease in fire injuries.

Harmathy adapted the percent reduction values in fire deaths and injuries and used the Ruegg/Fuller report to determine the number of deaths and injuries that would occur in sprinklered occupancies. However, his reference case for no sprinklers made use of 1985 NFIRS data and information from a fire protection consultant, therefore causing his results **to** differ from those presented by Ruegg/Fuller. According to Harmathy, there was a 75 percent decrease in life loss due to sprinkler and detector presence and a 46 percent decrease in fire injuries (27).

Waterborne Disease Outbreaks

Sobsey et al. reported that the total number of potentially pathogenic microorganisms in water is unknown and increasing (30). Pathogens in drinking water arc associated with acute risks that vary greatly in clinical manifestation and severity. The manifestation and severity are dependent on the microorganisms and the degree of exposure. Pathogens arc defined as either frank or conditional. Frank pathogens produce infectious disease in both healthy and compromised individuals, whereas, conditional pathogens produce disease only when conditions in the environment and the compromised host provide opportunity. Reliable detection methods do not currently exist for many of the pathogens commonly found in water. In addition, the mechanisms of pathogenicity are known for very few microorganisms and is virtually non-existent for many conditional pathogens (30). Microbial risk assessment modelling has indicated that exposure to low concentrations of microorganisms in water (one infectious microorganisms per 1,000 liters) may result in significant risks of infections, illness and mortality in a community. Sobsey et al. reported that the risk of microbial infection is commonly known to most individuals as nearly all children experience rotavirus and most Americans contract hepatitis A during their lifetime. Gastrointestinal illnesses am. on the average, experienced once or twice a year by most Americans (30).

Data on the occurrence of outbreaks of waterborne disease is collected and tabulated by the Centers for Disease Control in conjunction with the USEPA. Outbreaks arc reported on a voluntary basis and have been tracked since 1920. It is assumed that only a fraction of the total number of outbreaks are reported, however, the level of under reporting is not known. Source water is categorized into community water systems, noncommunity water systems and non-public or individual systems (typically individual wells). Community and noncommunity systems are classified as public systems and constitute 31 percent and 69 percent, respectively of the 189,600 public systems in the United States. Community systems service approximately 91 percent of the U.S. population (5).

Reported Outbreaks

The number of reported outbreaks associated with potable water for the years 1987, 1988, 1989 and 1990 were 15, 13, 12 and 14, respectively. The minimum number of reported outbreaks for each year from 1971 through 1986 was 20 (5). This data is presented in Figure 7.

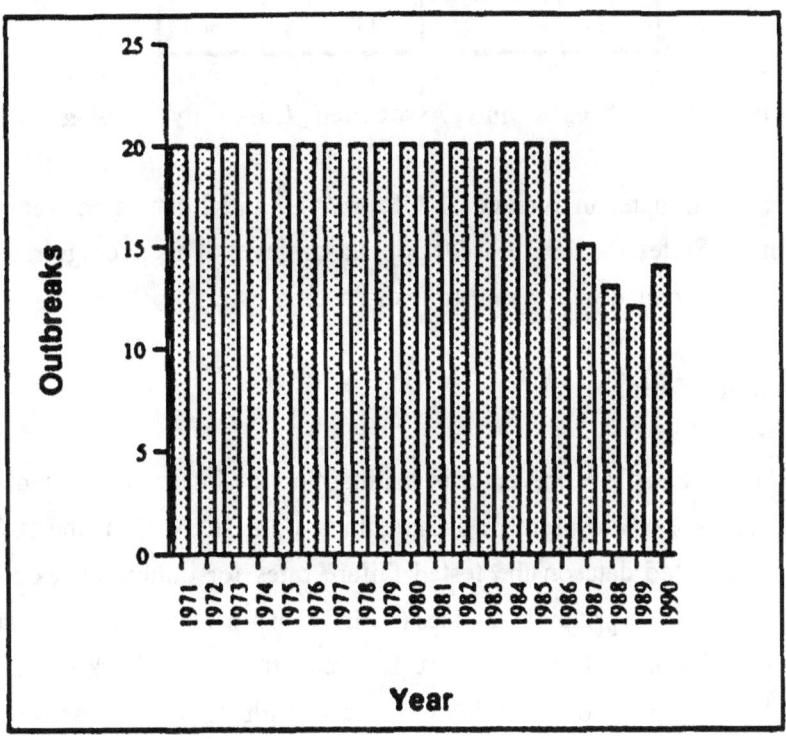

Figure 7 - Outbreaks of Waterborne Disease

Herwaldt et. al., recently studied the outbreaks of waterborne disease in the United States for the years 1989 and 1990 (5). For this two year period, 26 outbreaks of waterborne disease were reported and concluded to have resulted from the consumption of water intended for drinking. As a result of the outbreaks, illnesses inflicted 4,288 people and 4 people died. The number of outbreaks associated with community systems was 11, resulting in 1660 illnesses (39 percent of

29

the total reported illnesses) and four deaths (100 percent of total reported deaths). In the waterborne outbreaks caused by community water, several etiologic agents were identified; including Giardia lambio, Hepatitis A, Norwalk-like, and *E. coli.* Table 1 summarizes data repeated by Herwaldt et al. (5) in the U.S. for 1989-1990.

Etilogical Agent	outbreaks	Cases
AGI	4	894
Giardia	4	503
Hepatitis A		
Norwalk-like	0	0
E. coli 0157:H7	1	243
CLB (possible)	1	21
Total	11	1664

Table 1 - Number of Outbreaks and Associated Cases by Etiological Agent (5)

Based on this most recent data, an average of 830 illnesses and 2 deaths per year can be assumed to occur in the United States for the 1989-90 population level. This average rate will be used to identify present risks associated with waterborne disease.

Backflow Preventer Failures

The effectiveness of any backflow preventer in maintaining the integrity of potable drinking water is directly related to its expected rate of failure. In a field study (31). the AWWA - Pacific Northwest. Section reported data on the tested failure rates for double check valve assemblies (DCVA) and reduced pressure principle devices (RPBD). A breakdown of these data are shown in Tables 2 and 3, and in Figure 8. For this report, failure of the No. 1 check valve in the DCVA is used to estimate the probability of a single check valve failure. Such an assumption, of course, assumes that the single check valve device uses the same technology, or has an equal degree of reliability, as the first check valves placed in the DCVA. Failure of the DCVA and RPBD &vices is assumed to occur when both checks fail.

This breakdown of expected failure rates according to different backflow devices is presented because each device has distinct advantages and disadvantages. A SCV represents a low cost alternative to both the DCVA and the RPBD devices. A disadvantage of the SCV is that is has not been approved for low hazard or high hazard crossconnections. Based on information presented in this report, however, such a high degree of protection can not be justified for household sprinkler systems. The DCVA represents a higher cost backflow protection alternative and has the advantage of being approved for low hazard cross-connections. Another advantage of this device (and also the RPBD) is that it is supplied with test cocks and shut-off valves for periodic checking. The RPBD backflow preventer represents a high cost and high protection alternative (suitable for high hazard crossconnections). A distinct disadvantage of this backflow preventer is the potential for head loss, which may cause a sprinkler system to be ineffective, especially in retrofit situations.

Based on data printed in Tables 2 and 3, the average reported failure rate of DCVA's and RPBD and the estimated failure rate of a SCV in the 0.75 -to- 1.00 inch range is as follows:

DCVA - 1.6%

RPBD - 1.8%

SCV - 4.0%

The 0.75 -to 1.00 inch range is selected because it represents the typical household supply line. These failure ratings will be used in the next section to predict expected rates of risk when selecting each of these backflow protection devices. For example, if the risk of illness without any backflow protection device is 1.0×10^{-6} (one person out of one million will become ill), this risk will be reduced to 4.0% ($.04 \times 10^{-6}$ or 4.0×10^{-8}) if a SCV is used for protection.

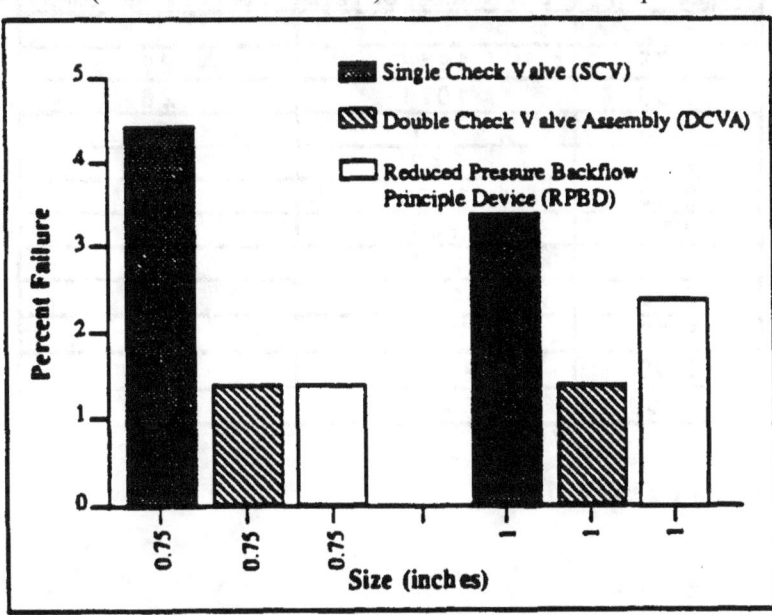

Figure 8 - Failure Rates for 0.75 and 1.00 inch Devices (31)

Size (in.)	Total Number of Tests	Single Check Valve Failed	Double Check Valve Failed
0.75	1465	65	22
1.00	1795	65	29
1.25	22	0	0
1.50	756	89	59
2.00	1155	67	59
2.50	54	5	4
3.00	213	15	17
4.00	307	12	12
6.00	226	8	6
8.00	272	17	12
10.00	67	7	8

Table 2 - Failures of DCVA and SCV (31)

Size (in.)	Total Number of Tests	Both Checks Failed
0.75	1946	29
1.00	1071	26
1.25	171	3
1.50	543	9
2.00	1158	19
2.50	146	1
3.00	252	3
4.00	273	3
6.00	158	3
8.00	62	0
10.00	28	0

Table 3 - Failures of RPBD (31)

32

SECTION 6.0: RISKS VERSUS BENEFITS

Introduction

Risk is considered the likelihood or probability that a particular event will have either a direct or indirect adverse effect on human health and welfare (32). Two risks are the subjects of this study: the adverse effects on human life/health due to fires, and the adverse effects on human illness due to waterborne disease. The overall objective of this study is to answer the question:

What is the difference in risk associated with a single check valve device, a double check valve assembly, and a reduced pressure principle device for residential sprinkler system?

This section explains how risk is quantified for this report, identifies conditions commonly assumed for both risk assessments, calculates risks associated with fire and water contaminations, and finally, compares these risks.

QUANTIFYING RISK

Risk is typically reported in either time or unit activity, such as worker days lost per work year due to illness or number of cancer cases per capita per year due to chemical exposure (32). Commonly used options for risk measuring units include number of lives lost per unit time or event, quality-adjusted life-years (QALYs), and economic burden (33). In evaluating risks associated with bacteria in drinking water supplies versus the risks associated with unsprinklered dwellings, a unit of risk was chosen. Units of risk for this paper arc number of lives lost per year, occurrences of fire injuries per year and number of waterborne illnesses per year. For example, a risk of 1×10^{-5} means that one person out of 100,000 will experience the cited event each year. The reader should note that this paper is comparing, and therefore equating, deaths and injury from fire to illness from unprotected water.

Common Assumptions

The piping and sprinkler system configurations presented in Section 3 were considered representative of typical one- and two-family dwelling in the United States. Water volume figures from household plumbing lines and sprinkler lines for this type of dwelling are used to estimate the increased amount of risk associated with sprinkler line installations. Such a direct correlation of risk increase to volume increase is based on the assumptions that intermittent water (household potable lines) have an equal probability of health risk to extended stagnant water (household sprinkler lines).

The total number of people living in one- and two- family dwellings was calculated from U.S. Census Bureau data from 1990 and is shown in Table 4.

Type of Dwelling	Number of Units
Single Unit	65,761,000
2 - 4 Units	9,876,000
Mobile homes and trailers	7,400,000
Subtotal	83,037,000
Total U.S. Dwellings	102.264.000 units

Table 4 - Classification of Dwellings

According to this table, 81.2 percent of the dwellings in the United States are one to four unit dwellings or mobile homes/trailers (83,037,000 divided by 102,264,000). A conservative assumption was made that all of these dwellings were one- and two-family dwellings and thus the subject of this risk assessment study. Using a total U.S. population for the year 1990 of 252,177,000, it can be estimated that 204,764,000 persons resided in single or double unit dwellings in the United States for the year 1990 (81.2% of the total population).

Waterborne Disease Risk Assessment

The USEPA has adopted a maximum acceptable risk for gastrointestinal illness caused by community water supplies of one illness per 10,000 people per year (a 104 annual risk). To ensure compliance with this risk level for infection from hepatitis A and most other enteric viruses, the total enterovirus concentrations in treated drinking water should be less than 2.2×10^{-7} per liter (9,34).

Experimental data pertaining to the concentrations or life expectancies of enterobacteria in stagnant lines could not be found during literature searches undertaken for this project Therefore, historical data was used to estimate the potential for waterborne disease occurrences. The following conditions were assumed in the development of waterborne disease risk assessment:

1) The reported waterborne disease values for the years 1989 and 1990 are for all outbreaks associated with community water systems and include deficiencies in both the treatment system and the distribution system (e.g., temporary disinfection interruption, inadequate filtration, cross-connection, back siphonage, etc.). Because the expected degree of risk from household sprinkler lines will be based on estimated risks already present from household potable lines, a conservative approach is to assume that a.U reported outbreaks in community systems were the result of contamination from household piping systems. In addition, this risk assessment also assumes that the degree of under reporting was 50%.

2) The increase in risk of contracting a waterborne illness from unprotected stagnant sprinkler lines was considered proportional to the ratio of potable water normally present in a residential dwelling to the combined volume of water from potable water plus sprinkler water, as modelled in Section 3. The estimated potable water system volume of water was 588 in3 and the stagnant sprinkler line volume of water was 832 in^3 or a 1:1.5 ratio. An increased risk of 2.5 is therefore assumed.

3) The number of reported illnesses and deaths for the year 1990 was used to represent the expected number of illnesses and deaths *without sprinkler* systems in one- and two-family dwellings.

4) It was assumed that **a** properly functioning backflow preventer (SCV, DCVA, or RPBD) would isolate the stagnant water associated with a sprinkler system, thereby eliminating the carbon source and the possibility of biofilm sloughing.

5) Due to the continual starting and stopping of water in household piping systems, it was assumed that sloughing of harmful biofilms could occur readily.in a .potable water line. It was also assumed that this same phenomenon would occur in the total length of in an unprotected sprinkler water line (e.g. complete intermixing).

35

6) Biofilm growths are characterized by huge initial occurrences of coliform organisms followed by a gradual decline in bacterial levels over time, with the period of time between the large initial occurrences and the gradual decline lasting several months (12). However, for the purposes of this risk assessment it was conservatively assumed that the stagnant water in sprinkler lines would be at the peak of harmful bacteria concentrations during any introduction to potable water lines. Possible die-off of bacteria in stagnant sprinkler lines, therefore, is not considered.

Estimated Risk of Waterborne Illness

The following equation was used to estimate the increase in illnesses/deaths that could potentially occur if sprinklers without a single check valve or backflow devices were installed throughout the United States.

$$R = N * U / P$$

where:

R   = the risk of illness/death due to waterborne diseases

N   = number of reported illnesses and deaths for a year or averaged years for which data exists

U   = degree of under reporting

P   = population of U.S. citizens living in one and two-family dwelling in data Year.

For the years 1989-1990, an average of 830 illnesses and 2 deaths per year (N) were reported. This number is increased according to the assumed degree of under repotting. Using a total of 204,764,000 people assumed to be living in one- and two-family dwellings (P), the probability of contracting an illness due to waterborne disease in those years was:

$$R \quad = \quad 830 * 2/204,764,000$$
$$R \quad = \quad 8.2 \times 10^{-6}.$$

In other words, for every one million people (10), 8.2 people will become ill.

To estimate the expected degree of risk associated with the addition of household sprinkler lines, the following scenarios were considered:

Scenario A: No protection device (SCV, DCVA, RPBD) is provided.

Scenario B: Either of the above three devices (SCV, DCVA, RPBD) are placed at the sprinkler line.

Scenario C: Either of the above three devices (SCV, DCVA, RPBD) are placed at the sprinkler feed line and the potable water feed line.

Scenario D: Same as Scenario C, but the device is placed at the feed line that branches off to both the potable feed line and the sprinkler feed line.

Figure 9 shows a schematic for each of these scenarios.

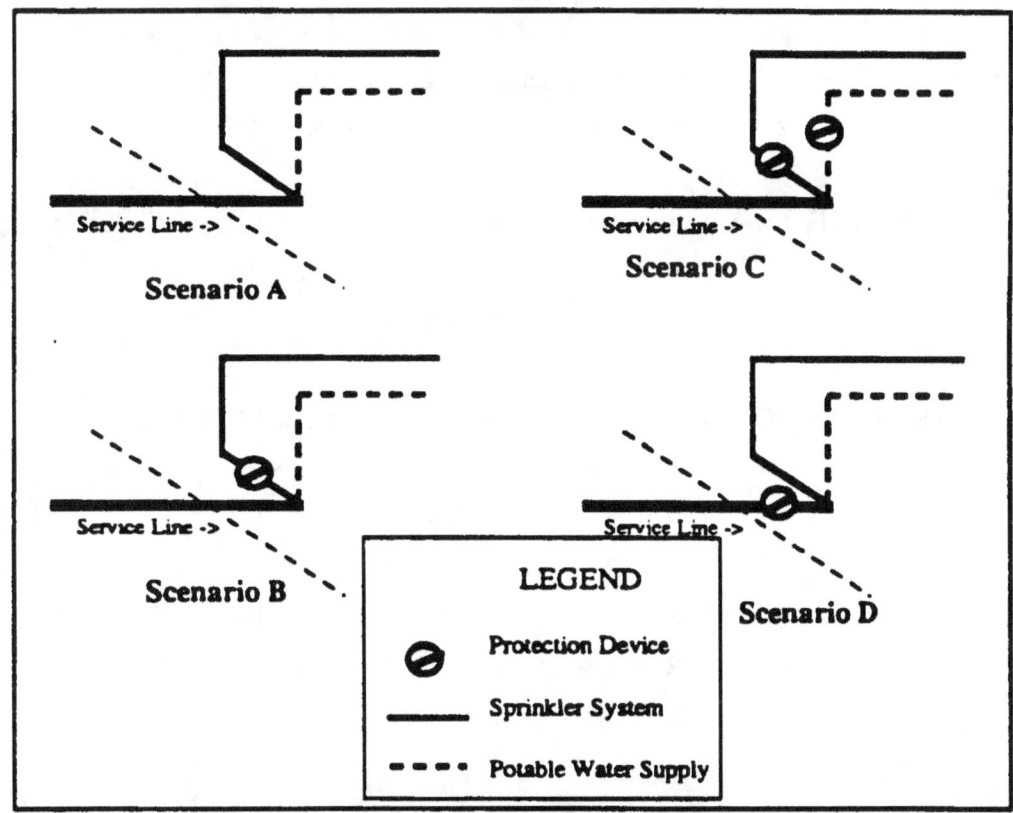

Figure 9: Location of Sprinkler and Water Line Protection Devices

The risk associated with Scenario A is found by multiplying 8.2×10^{-6} by a factor of 2.5 (the total increase of water volume resulting from adding a sprinkler line - see assumption 2).

$$\text{Scenario A: } R = 8.2 \times 10^{-6} \times 2.5$$
$$= 2.05 \times 10^{-5}$$

It should be noted that this risk level is an order of magnitude lower than the level of risk that is considered acceptable by the USEPA for occurrences of gastrointestinal illnesses.

For Scenario B, risk is calculated by adding the degree of risk associated with a household's potable water line to the increased risk of adding a protected sprinkler line.

$$\text{Scenario B: } R = 8.2 \times 10^{-6} + R \text{ (protected sprinkler line)}$$
$$\text{where: } R \text{ (proucted sprinkler line)}$$
$$= 8.2 \times 10^{-6} \times 1.5 \times \% \text{ failure}$$

The 1.5 factor accounts for an increased water volume associated with a sprinkler system while the % failure relates to whatever protection device is selected SCV. DCVA, RPBD).

$$\text{Scenario B (SCV): } R = 8.2 \times 10^{-6} + (8.2 \times 10^{-6} \times 1.5 \times 0.04)$$
$$R = 8.2 \times 10^{-6} + 4.92 \times 10^{-7}$$
$$R = 8.70 \times 10^{-6}$$

$$\text{Scenario B (DCVA): } R = 8.2 \times 10^{-6} + (8.2 \times 10^{-6} \times 1.5 \times 0.016)$$
$$R = 8.2 \times 10^{-6} + 1.97 \times 10^{-7}$$
$$R = 8.40 \times 10^{-6}$$

$$\text{Scenario B (RPBD): } R = 8.2 \times 10^{-6} + (8.2 \times 10^{-6} \times 1.5 \times 0.018)$$
$$R = 8.2 \times 10^{-6} + 2.21 \times 10^{-7}$$
$$R = 8.42 \times 10^{-6}$$

As expected. the degree of risk associated with installing any type of backflow preventer (SCV, DCVA, RPBD) at the sprinkler line alone is close to the degree of risk initially calculated to be present with a typical household water supply system. Also, as shown from these calculations, the difference in risk between these devices is not great. If total protection from household water backflow is to be realized, however, a backflow device should be installed at the potable water supply line as well as the sprinkler supply line (Scenarios C & D).

Risk calculations for Scenarios C & D are the same if risk within the individual household is not considered. In other words, if the Scenario C option is chosen, backflow from the sprinkler line to the residential potable line is prevented. The Scenario D option does not provide this protection within the household, but does provide the same amount of protection to the surrounding community. Of course, if all homes in the community are quipped with a sprinkler system, the option D Scenario cannot be considered equal to the option C Scenario. Expected risks for Scenarios C & D are calculated as follows:

$$\text{Scenario C or D: } R = 2.05 \times 10^{-5} \times \% \text{ failure}$$

The 2.05×10-s factor accounts for degree of risk associated with a sprinkler system plus the potable water lines (Scenario A), while the 8 failure relates to whatever protection device is selected (XV, DCVA, RPBD).

Scenario C or D (XV):	R	=	$2.05 \times 10^{-5} \times 0.04$
	R	=	8.20×10^{-7}
Sceriario C cm D (DCVA):	R	=	$2.05 \times 10^{-5} \times 0.016$
	R	=	3.28×10^{-7}
Scenario C or D (RPBD):	R	=	$2.05 \times 10^{-5} \times 0.018$
	R	=	3.69×10^{-7}

A comparison of these options to the Scenario B option shows an increase in protection if a backflow device is installed for the household supply line as well as the sprinkler line rather than simply at the sprinkler supply system. Again, as shown in the option B calculations, the difference in protection from these devices (XV, DCVA, RPBD) is very slight

Fire-related Deaths/Injuries Risk Assessment

The Ruegg/Fuller data presented in Section 5 was based on the following assumptions:

1) The base case for calculating the number of deaths and injuries resulting from fire without sprinkler systems assumes that all homes are equipped with a smoke detector with an associated reliability of 84 percent;

2) The calculation for number of deaths/injuries prevented assumes that all homes are quipped with sprinklers and smoke detectors with associated reliability of 92 and 84 percent respectively.

Estimated Risk of Fire Death/Injury

Using the Ruegg/Fuller data presented in Section 5, which had the more conservative of the results, and applying it to the year 1990, 2120 lives would have been saved and 6700 injuries would have been prevented with the widespread application of residential sprinkler systems. This data was based on 3370 civilian deaths and 15250 civilian injuries in 1990.

$$\text{Saved Lives} = 3{,}370 \times .63$$
$$= 2.120$$

$$\text{Saved Injuries} = 15250 \times 0.44$$
$$= 6{,}700$$

Assuming, as before, that 204,764,000 people lived in one and two family dwellings, the present risk associated with death and injury if homes are not equipped with sprinkler devices are calculated as follows:

$$\text{Lives:} \quad R = 3{,}370 \, / \, 204{,}764{,}000$$
$$= 1.65 \times 10^{-5}$$

$$\text{Injuries:} \quad R = 15250 \, / \, 204{,}764{,}000$$
$$= 7.45 \times 10^{-5}$$

Estimated risks if sprinkler systems are universally installed are calculated as follows:

$$\text{Lives:} \quad R \quad = \quad (3,370 - 2,120) / 204,764,000$$
$$= \quad 6.10 \times 10^{-6}$$

$$\text{Injuries:} \quad R \quad = \quad (15,250 - 6,700)/204,764,000$$
$$= \quad 4.18 \times 10\text{-}S$$

A Comparison of Risks (Fire vs Illness)

A summary of the previous risk assessments is given in Tables 5 and 6. As can be seen from Table 6, the relative risk associated with fire events (death/injury) is 11.10 compared to a 1.00 risk associated with a waterborne illness event with the present assumed condition of no sprinkler systems. The level of risk associated with an unsprinklered household is expressed in terms of probable death (1.65×10^{-5}) and probable injury (7.45×10^{-5}). while the level of risk associated with health impacts from unprotected sprinkler lines is expressed in terms of probable cases of gastrointestinal illnesses (2.05×10^{-5}). It is important to note that the 2.05×10^{-5} risk is still below the 1.0×10^{-4} USEPA target risk level.

A relative reduction in fire related risk events to 5.84 (from the original 11.10) shows that the residential sprinkler systems would have a significant effect in preventing fire-related deaths and injuries. Even without any reverse flow prevention (Scenario A assumes no device is installed), the increased relative risk of waterborne illnesses (2.50) is still below the reduced risk of fire death and injury (5.84).

Scenarios B and C/D illustrate the reduced risk of waterborne illness associated with the selection of either a SCV, DCVA. or RPBD. As can be seen from both Tables 5 and 6, very little difference in risk is noted between either device. Residential sprinkler systems equipped with a single check valve device with the same degree of reliability as detected in a more expensive double check valve assembly, therefore, should be as effective as the more expensive DCVA and RPBD backflow preventers.

The Scenario C/D approach is included in this risk assessment study to illustrate the degree of risk reduction (almost ten fold from the Scenario B approach) that should be realized by also installing reverse flow devices at the potable water connection as well as the sprinkler water connection.

41

Condition	Death	Fin Event Injury	Combined	Waterborne Event (Illness)
Household Water Only (No Sprinkler)	1.65×10^{-5}	7.45×10^{-5}	9.10×10^{-5}	8.2×10^{-6}
Option A	6.10×10^{-6}	4.18×10^{-5}	4.78×10^{-5}	2.05×10^{-5}
Option B	6.10×10^{-6}	4.18×10^{-5}	4.78×10^{-5}	8.70×10^{-6}
SCV	↓	↓	↓	8.40×10^{-6}
DCVA				8.42×10^{-6}
RPBD				
Options C& D	6.10×10^{-6}	4.18×10^{-5}	4.78×10^{-5}	8.20×10^{-7}
SCV	↓	↓	↓	3.28×10^{-7}
DCVA				3.69×10^{-7}
RPBD				

Table 5: Units of Risk for Fire (Injury/Death) and Waterborne Illness Events

Condition	Fire Event (Injury & Death)	Waterborne Event (Illness)
Household Water Only (No Sprinkler)	11.10	1.00
Option A	5.84	2.50
Option B	5.84	
S C V		1.06
RPBD		1.03
Options C& D	5.84	
SCV		0.10
DCVA		0.04
RPBD		0.05

Table 6: Normalized Risk of Fire (Injury & Death) and Waterborne Illness Events

SECTION 7.0: AREAS NEEDING ADDITIONAL INFORMATION

Although conclusions presented in the next section of this report am based on rational evaluations of available information, this literature study had to make certain assumptions. These assumptions and the need for further study are addressed below.

A Single Check Valve Device (SCV)

For this study, a SCV was assumed equal in performance reliability to the type of check valve found in a DCVA (the No. 1 check valve). Such an assumption could not be based on field data as that type of data collection was not found. Backflow preventer evaluations are naturally concerned with DCVA and RPBD devices. It is unlikely that the location and performance reliability of SCV's is even known. Information regarding the reliability of a low cost single check valve device, therefore, should be generated form field surveys or laboratory experiments.

Biological Impacts

Experimental and field related data on the effects of stagnation in piping systems has been limited to assessments over periods that may typically occur in potable water supply systems. This report refers to these conditions as intermittent stagnation, where intermittent flow or very low velocity flow does occur. Intermittent flow conditions will transport needed nutrients and oxygen for biological growth and will cause biofilm materials to slough off. Extended stagnation periods, where water will remain in the pipeline for periods of a year or more, may be very different. Under such conditions, needed nutrients, oxygen, and other trace materials may not be transferred to the water, causing bacterial die-off or very slow bacterial growth rates. A simple extrapolation of intermittent stagnation effects, therefore, may not be appropriate. Ideally, the following information should be available:

- What is the degree of mixing between an unprotected (or failed protection device) extended stagnation line with an intermittent stagnation line as typically found within a household potable line/sprinkler line connection.

- What is the degree of material diffusion (nutrients, oxygen, etc) for this type of connection.

43

- What are the life cycles, under extended stagnant conditions, for a variety of bacteria typically expected in a potable water network.

- What is the level of risk contracting an illness associated with waterborne disease that is caused by typical starting and stopping of water in household potable water lines relative to the risk associated with extended stagnation in sprinkler systems.

Chemical Impacts

As with biological impact evaluations, chemical impacts resulting from internal pipe corrosion reactions has been limited to assessments over periods that may typically occur in potable water supply systems. The literature does describe expected copper corrosion reactions that occur under various conditions such as with oxygen or without oxygen, but field data describing expected corrosion during extended stagnation was not found. This type of data is not expected to be generated from experiments that are currently being required to meet the new lead/copper regulations (10), as only intermittent stagnation conditions will be simulated. Information regarding expected water quality impacts from PVC pipe materials was also not found in the literature. Ideally, the following information should be available:

- How do extended stagnation corrosion conditions compare to intermittent stagnation conditions for copper pipes suitable for potable water supplies. Parameters of this evaluation should include chemical variations, including oxygen, carbon dioxide, pH, and conductivity, as well as the physical variation of flow.

- Possible internal corrosion reactions that will result from the introduction of various lead/copper control strategies should be evaluated for extended stagnation conditions. These strategies should include recommended practices of flushing, pH adjustments, and various pipe coating approaches.

- Impacts on water quality from extended stagnation in PVC pipelines (and various recommended solvents) should be evaluated from field or laboratory studies.

Physical Impacts

Stagnant water in a pipe system will contribute to turbidity in the water supply if intermittent flow conditions resuspend particles that either collected or were generated from tuberculation. As with biological and chemical impacts, the potential degree of physical impact from extended stagnation dependents on the degree of intermixing at the connection location between the potable pipe network and the sprinkler pipeline. Variables that should be included in the evaluation of inter-mixing include flow velocities and connection configurations.

SECTION 8.0: RECOMMENDATIONS

Assessments presented in this report illustrate that the risk of death and injury associated with unsprinklered dwellings is higher that the risk of illness associated with unprotected sprinkler systems. This assessment alone supports the need for residential sprinkler systems.

Calculations presented in this report also show that the three protection devices evaluated in this study (a single check valve device, a double check valve device, and a reduced pressure principle device) provide the same degree of health risk protection. Either device will adequately protect human health to orders of magnitude below the desired USEPA acceptable risk level of 1×10^{-4} gastrointestinal illnesses per year per person.

Recommendations of this report are as follows:

1) The SCV device should be considered in favor of the DCVA and the RPBD for individual household sprinkler systems.

2) If a protection device is to be used for a sprinkler line, the same degree of protection should be considered at the potable water line.

3) Areas requiring additional study (as presented in the previous section) should be seriously considered.

References

1.   Karter, MJ. Jr., "Reports on U.S. Fire LOSS," <u>NFPA Journal,</u> (September/October 1992): 36-43.

2.   Ruegg, R.T., Fuller, SK., "A Benefit Cost Model of Residential Fire Sprinkler Systems", NBS Technical Note 1203, Operations Research Division, Center of Applied Mathematics, National Engineering Laboratory, National Bureau of Standards, Gaithersburg, Maryland, November 1984;

3 . Regli, S., Rose, J.B., Haas, C.N., and Gerba, C-P., "Modeling the Risk from Giardia and Viruses in Drinking Water", <u>J. AWWA,</u> 10/91, pp. 76-84.

4 . Seidler, RJ. and Evans, T.M., Persistence and Detection of Coliforms in Turbid Finished Drinking Water", USEPA Publication No. EPA-600/82-82654, August 1982.

5 . Henvaldt, B-L., Craw, G.F., Stokes, S.L. and Juranek, D.D., "Outbreaks of Waterborne Disease in the United States: 198990." <u>J. AWWA,</u> pp. 129-135, April 1992.

6 . LeChevallier, M.W., "Microbial Processes within the Distribution System," American Water Works Association Conference Proceedings. 365-376 (1991).

7.   "Control of Biofilm Growth in Drinking Water Distribution Systems", EPA/625/R-92/001, June 1992.

8. LeChevallier. M.W..Schultz W.. Lee. R.G.. "Bacterial Nutrients in Drinking Water," <u>Applied and Environmental Microbiology.</u> pp. 857-862, March 1991.

9. Pontius., F.W. , "A Cement Look at the Federal Drinking Water Regulations," <u>J. AWWA,</u> pp. 3650, March 1992.

10 . "The Lead and Copper Rule," <u>Journal AWWA</u> (July 1991): 12.

11. Ripka, L.V.. <u>Plumbing Installation and Design,</u> American Technical Society, Chicago, 1978

12. LeChevallier, M. W., "Coliform Regrowth in Drinking Water: A Review," J.AWWA, 82(11):74-86 (1990).

13. Ibrahim, AA. and M. Sadiq, "Metal Contamination of Drinking Water from Corrosion of Distribution Pipes," Environmental Pollution, 578 (1989), 167-78.

14. Wolfaardt, G. M., Cloete, T.E., "The Effect of Some Environmental Parameters on Surface Colonization by Microorganisms," Water Resources. Vol. 26, No. 4, pp. 527-537, 1992.

15. Emde, K.M.E., Smith, D.W., Facey, R, "Initial Investigation of Microbially Influenced Corrosion (MIC) in a Low Temperature Water Distribution System," Water Resources. 26(2): 169-175 (1992).

16. LeChevallier, M.W., Babcock, T.M., Lee, R.G., "Examination and Characterization of Distribution System Biofilms," Applied and Environmental Microbiology. pp. 2714-2724, December 1987.

17. Donlan, R.M. and Pipes, W.O., "Selected Drinking Water Characteristics and Attached Microbial Population Density," J.AWWA, pp. 70-76, November 1988.

18. Wagner, J. and W.T. Young, "Corrosion in Building Water Systems," J. Environment Treatment & Control, October 1990,40-6.

19. Leidheiser, H, The Corrosion of Copper, Tin and their Alloys, Robert E. Krieger Publishing Co, Huntington, NY 1979.

20. Mendanhall, J.H., Understanding Copper Alloys, Robert E. Krieger Publishing Co, Malabar, FL 1986.

21. Alleman. J.E., and HE. Hickey, "Investigation of the Water Quality and Condition of Pipe in Existing Automatic Sprinkler Systems for the Analysis of Design Options with Residential Sprinkler Systems," NBS-GCR-82-399, August, 1982.

22. "Fire Loss in the United States", Fire Journal, September issues for 1978-1990.

23. "Fire Loss in the United States", Fire Journal, September 1991, p. 48

24. Karter, M.J., Jr. "Fire Loss in the United States During 1991". NFPA, Fire Analysis and Research Division, Quincy, MA, September 1992, p. 7.

25. Melinek, S.J., "Potential Value of Sprinklers in Reducing Fire Casualties", Fire Safety Journal 20 (1993) : 275-287.

26. "Backflow Protection for Fire Sprinkler Systems," NFSA, R.R. No. 17054.

27. Harmathy, T.Z., "On the Economics of Mandatory Sprinklering of Dwellings", Fire Technology, August, 1988, pp. 245-61.

28. "Fire Protection Handbook", Seventeenth Edition, National Fire Protection Agency, pp. 5-190.

29. Unpublished report by A. Gomberg et al, "A Decision Model for Evaluating Residential Fire Risk Reduction Alternatives", NBSIR, National Bureau of Standards, in preparation, 1984.

30. Sobsey, M.D., Dufour, A-P.. Gerba, C.P., LeChevallier, M.W. and Payment, P., "Using a Conceptual Framework for Assessing Risks to Health from Microbes in Drinking Water," J.AWWA, pp 44-48, March 1993.

31. American Water Works Association - Pacific Northwest Section, Cross Connection Control Committee, "Summary of Annual Test Reports Reduced Pressure Principle and Double Check Valve Assemblies, December 1988.

32. Wentz, CA., Hazardous Waste Management,. McGraw Hill Book Co., New York, 1989.

33. Putnam, S.W.. Graham, J-D.. "Chemicals versus Microbials in Drinking Water: A Decisions Sciences Perspective," J.AWWA. pp. 57-61, March 1993.

34. Regli. S., Rose, J.B.. Haas, C.N., and Gerba, C-P., "Modeling the Risk from Giardia and Viruses in Drinking Water," J.AWWA, pp. 76-84, November 1991.

49

35. Olson, B.H. "Assessment and Implications of Bacterial Regrowth in Water Distribution Systems," U.S. Environmental Protection Agency Publication Number EPA-600/S2-82-072, September 1982.

36. Seidler, R.J. and Evans, T-M., "Persistence and Detection of Coliforms in Turbid Finished Drinking Water," U.S. Environmental Protection Agency Publication Number EPA-600/S2-82-054, August 1982.

37. Gaudy, A.F., Jr., Gaudy, E.T., <u>Microbiology for Environmental Scientists and Engineers.</u> McGraw-Hill Book Company, pp. 666-682,198?.

38. LeChevallier, M.W., Cawthon, C.D., Lee, R.G., "Factors Promoting Survival of Bacteria in Chlorinated Water Supplies," <u>Applied and Environmental Microbiology.</u> pp. 649-654, March 1988.

39. "The Recommended Practice for Backflow Prevention and Cross-connection Control", American Water Works Association M-14 Manual, (1990). p. 25.

40. "Potential Value of Sprinklers in Reducing Fire Casualties", <u>Fire Safety Journal,</u> (20) 1993, pp. 275-287.

Appendix 1: Design Specifications for Typical Household Potable Piping Systems

The following design specifications were used to design a typical household piping system. The resultant example design is illustrated in Figure 1.

- 3/4-inch pipe is the minimum size water service for any building from the street to the water main.

- 3/4-inch pipe is the minimum size of building supply pipe (the fast section of water distribution piping within the building).

- 3/4inch pipe is the minimum size to a sill cock or lawn faucet.

- 3/4-inch pipe is the minimum size cold water supply to **a** water heater.

- 3/4-inch pipe is the minimum size for the first section of hot water pipe on the outlet side of a water heater.

- 1/2-inch is the smallest size for concealed pipe.

- **A** maximum of three fixtures in the same bathroom of a house may be supplied by the same 1/2-inch pipe.

- 1/2-inch is the minimum size fixture branch pipe, with the exception of 3/4-inch minimum size fixture branch pipe for the sill cock, urinal(w/ flushometer valve), and sinks and a l-inch minimum branch pipe for a water closet(w/ flushometer).

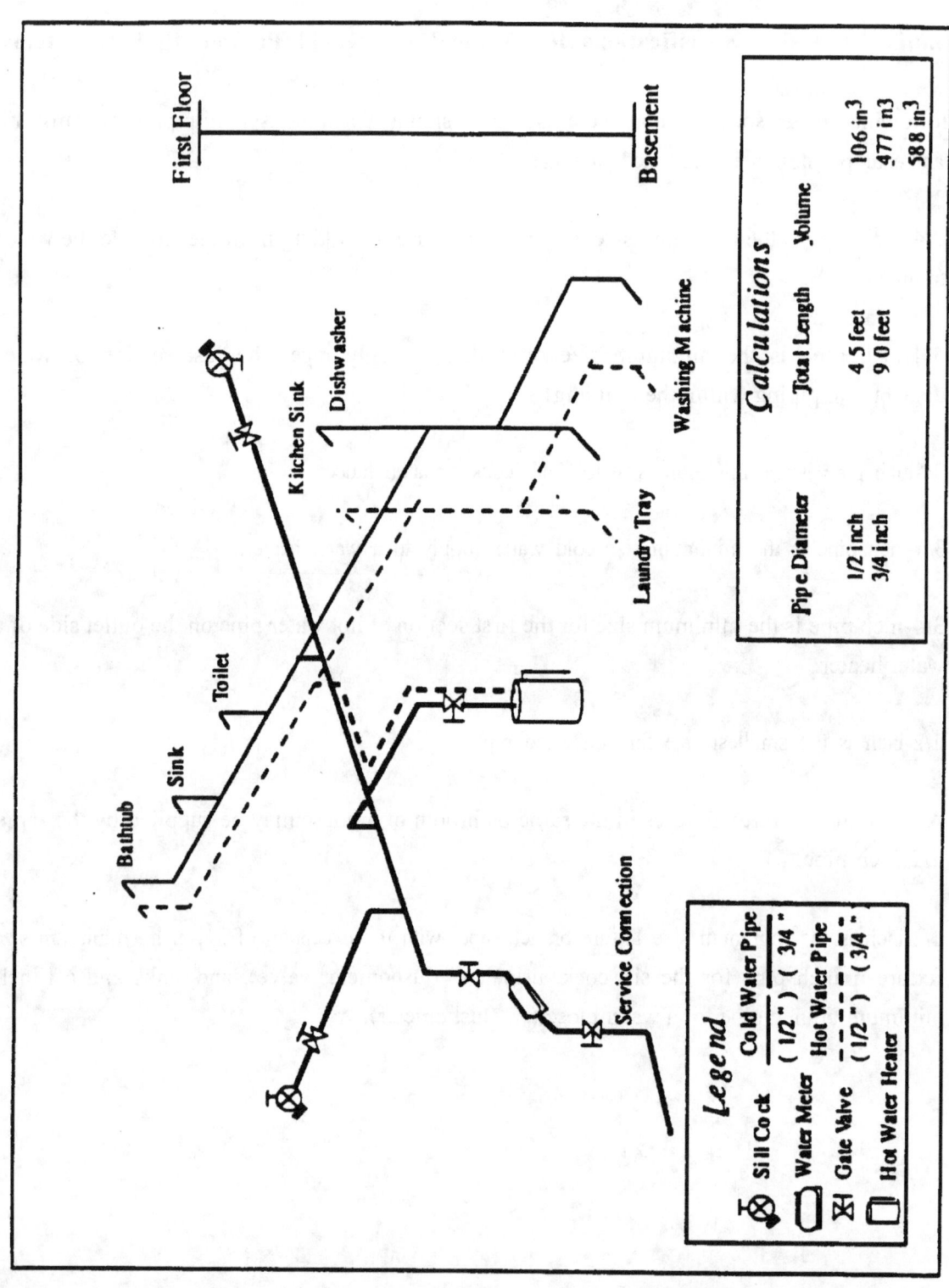

Figure 1: Typical Hot and Cold Piping System for a Single Family Dwelling

Appendix 2: Design Specifications for a Typical Household Sprinkler System

The following designs specifications were used to model an example household sprinkler system. The resultant design is illustrated in Figure 2.

- flow requirement of at least 18 gpm is required for one head activating at the hydraulically most remote location;

- flow requirement of at least 13 gpm is required for two heads activating at the hydraulically most remote location;
- maximum design area of 144 sqft per head;

- maximum spacing of 12' between heads; and

- maximum spacing of 6' off of wall.

The hydraulic verification of a typical sprinkler system was performed using a commercially available computer modeling program (Turbofire Z). Boundary conditions of the water supply were assumed as follows:

- 60 psi residual pressure; and
- 40-foot length of 3/4-inch service pipe line from main to house.

The hydraulic flow testing results are presented in Table 7.

	Required Flow (minimum)	Design Flow (actual)
1 Sprinkler Head @ Most Remote Location	18gpm	23.8 gpm
2 Sprinkler Heads @ Most Remote Location	13 gpm	1st @ 13.7 gpm 2nd @ 14.6 gpm

Table 7: Hydraulic Model Flow Testing Results

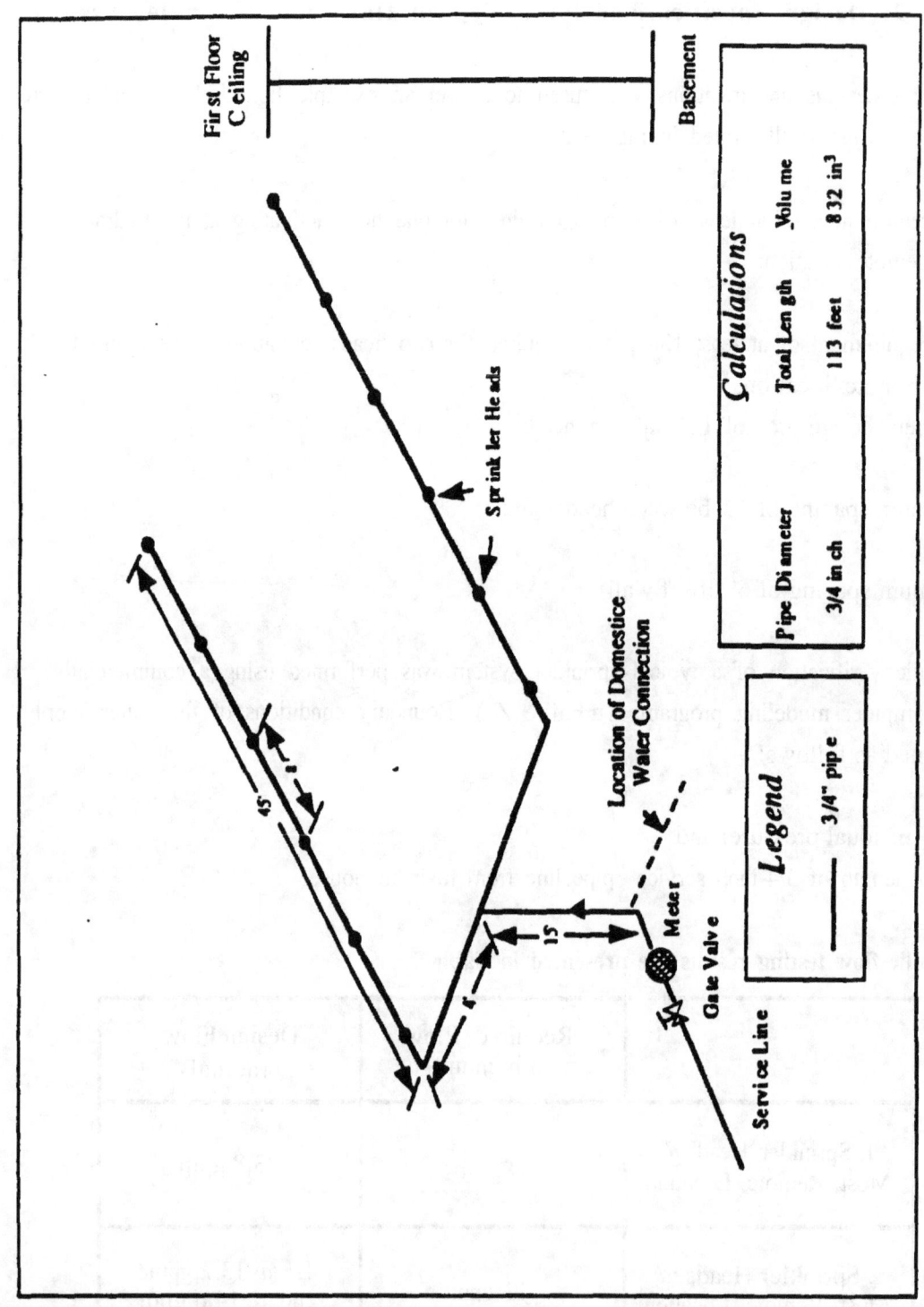

Figure 2: Typical Sprinkler System for a Single Family Dwelling

www.ingramcontent.com/pod-product-compliance
Lightning Source LLC
Chambersburg PA
CBHW081225170526
45165CB00009B/2951